How To Grow Your Very Own Fruit Trees

Quick Start Guide

HTeBooks

Disclaimer

This book is designed to provide condensed information. It is not intended to reprint all the information that is otherwise available, but instead to complement, amplify and supplement other texts. You are urged to read all the available material, learn as much as possible and tailor the information to your individual needs.

Every effort has been made to make this book as complete and as accurate as possible. However, there may be mistakes, both typographical and in content. Therefore, this text should be used only as a general guide and not as the ultimate source of information. The purpose of this book is to educate.

The author or the publisher shall have neither liability nor responsibility to any person or entity with respect to any loss or damage caused, or alleged to have been caused, directly or indirectly, by the information contained in this book.

Table of Contents

How Will This Book Help You?

Fruits are a store of goodness. They are tasty, juicy and nutritious, and they are packed with essential vitamins for your health and well being. It is therefore, a pleasureable responsibility to plant fruit trees. People have planted fruit trees all over the world, since the early times. Some people even show interest in growing fruit trees when they only have a limited space in their compounds. However, if you do not have the necessary knowledge and skills to plant and maintain your fruit trees, they will eventually die, or at the very best, become less productive. This means that all the effort you put into site preparation, planting and any other input will go to waste. In this book, you will learn how to grow and care for your fruit trees, as well as determine the right type to plant for your specific location in order to get the best production with minimum work.

Caveat: Very few homes can comfortably accomodate a real orchard, and neither would they want to. Growing fruit trees in large scale can be pretty labor intensive. However, one or two fruit trees can be a great addition to your home landscape.

Site Selection

So you have decided to jump into the fruit growing bandwagon and are ready to get dirty. However, before things get messy, a little knowledge and planning will go a long way. Below are some critical factors that you need to consider as you choose a suitable site for planting trees.

Adequate Sunlight

If you are looking to start your own fruit tree garden, it is important to note that most fruit trees thrive well under eight hours of sunshine every day, so your first priority should be to identify a planting site that receives plenty of sunlight. The sun in the early morning dries off dew from the foliage quickly and limits onset of diseases, while the midday sun improves the fruit flavor. In addition, you should ensure that your site selection does not fall under tree shades in order for your fruit trees to receive adequate sunlight and to avoid root competition.

Proper air drainage

There should be good air drainage as well, so avoid low locations and areas enclosed by shade trees or buildings where cold air resides. The north face of a building or a north-facing slope is an ideal location for growing frost sensitive plants such as apricots, sweet cherries, Japanese plums, and peaches. The shadow cast by the building or slope during winter makes the plant cooler, and this subsequently delays blooming in spring and bud development. On the other hand, the higher angle of the summer sun provides

5

enough sunlight in the growing season. A north-facing slope will also delay fruit ripening. A south-facing wall or slope will speed up both harvest and blooming, so you will need to provide extra protection against winter sunscald and frosts. If you live in an area that has a regularly cloudy or short growing season, then you may benefit more from planting the hardier fruits on a south facing wall or slope, as this tends to receive more intense sunlight. Good examples of hardier fruits include tart cherries, pears, apples, and American or European hybrid plums.

If the summers in your area are frequently very hot and dry, avoid planting in the south or southwestern slopes, or if that is not possible, be sure to irrigate the fruit trees thoroughly on a regular basis. Slopes facing either east or west have intermediate effects. If your area is frequented by strong winds, find a site that is shielded by existing buildings or plantings, or simply plant a windbreak. Windbreak trees in particular, can serve as screens or border plantings, provide nesting sites and shelter for insect eating birds, and even attract pest birds away from your garden.

Soil type and texture

Apart from sunlight and air considerations, soil type is also a major factor to observe when choosing a planting site. When you plant a fruit tree, chances are it will remain in the same exact location for several years, so it is vital to examine and prepare the soil diligently before planting. Even if you have to delay the planting project by a year, the increased health and tree growth will more than cover for that year. You can obtain valuable soil survey information from your local extension service agent.

Fruit trees thrive well in well-drained soil, therefore avoid planting in low areas that are prone to water puddles during heavy rains. If there is no well-drained site in your area, you can build raised beds or mounds, or even install drainage tiles where you want to plant your fruit trees. To prepare mounds, combine equal parts of compost and soil from other parts of your yard. You will also have to build raised beds for your soil if it is not deep enough for the roots of the trees. If you think that the soil at your site is shallow, use a soil auger or dig a hole to the necessary depth. As you dig, look for a sudden change in soil texture. Look out for the soil that becomes abruptly clayey – this soil type has a tendency to restrict root growth and water drainage into the soil below. While proper soil drainage is an important consideration for growing fruit trees, their growth will also be influenced by the soil texture. Sandy soils, clay loams or loam soils are better than heavy clay or very sandy soils when it comes to fruit trees. However, all fruit tree rootstocks have different adaptations. Some rootstocks endure sandy soils while others withstand clayey soils. Generally, apricots, Japanese plums, sweet cherries, and peaches thrive in sandy soils, while European plums, tart cherries, pears, and apples do well in heavier soils.

Soil pH

The pH of the soil is also very important when you want to plant fruit trees. The best approach to determine your soil's pH is to have your soil tested, possibly one season before you decide to plant your fruit trees. Add sulphur or lime if necessary in order to adjust the pH to between 5.5 and 7.0, which is ideal for most fruit trees. If potassium and phosphorus levels are low, add potash or phosphate fertilizers. In addition, fruit trees tend to thrive well in soils that have a high concentration of organic matter. Therefore, it may come in handy to turn under one or two green manure crops before you plant.

Pests and diseases

Plant fruit trees that are susceptible to nematode where there has been a lawn or grass cover crop growing for the last two years. Examples of these crops include apricots, cherries, plums, and peaches. Two years of no broad leaved plants is enough to almost completely get rid of troublesome nematodes, lesion and root knot, not to mention verticillium wilt. Verticullium or nematode infected fruit trees tend to be stunted and fruitless, and there is no cure once they are infected.

Avoid planting fruit trees where vulnerable plants have grown in the past two years. If for instance, you have to plant stone fruits in an area where vegetables grew recently, find a spot where corn has grown. However, before you plant, have a nematode test for your soil. If high levels of lesion or root knot nematodes are discovered, plant a trap crop of marigold on the ground the season before planting as these will lure nematodes to their roots. Go for a marigold variety that has been proven to serve as a good trap crop, for instance nemagold or tangerine blend. Let them stand at 6 to 7 inches apart in order to create a solid stand, and be sure to weed them. Pull out the marigold in early fall, along with the roots, and then destroy them. Plant a rye winter cover crop to further protect the soil from nematodes, then till it in before planting the trees early next spring.

It's fine to plant your fruit trees in a lawn, as long as you remove the sod 3 to 4 feet in a circle away from each fruit tree, and blend some organic matter such as compost into the whole area. Have a nematode test done if the lawn had several broad-leaved weeds. Get rid of any harmful perennial weeds like quack grass, and eliminate any competitive grasses like zoysia or Bermuda from the whole area that your fruit trees will eventually occupy. Instead, replace these with groundcovers that are less competitive. However, avoid replacing an old fruit with a closely related or similar type of fruit. These are probably infested with nematodes and several other soil borne diseases. Moreover, the roots of some fruit trees like peaches excrete a toxin that hinders the growth of budding peach roots.

*Key point/action step

Site selection is the first step you need to go through in order to have the most suitable site for your fruit trees. Remember that different fruit trees will require different sites; hence, use the information obtained in this chapter to determine the perfect site for whatever fruit trees you want to grow.

Plant Selection

"Never cut a tree down in the wintertime. Never make a negative decision in the low time. Never make your most important decisions when you are in your worst moods. Wait. Be patient. The storm will pass. The spring will come."

- Rober H. Schuller

Once you have determined and established your planting site, the next step is the most fun – finding the right fruit trees to plant in your garden. Depending on the nature of your location and the availability of natural resources, you will find that there are several types of fruit trees to choose from. Generally, there are three major types of fruit trees: dwarf, semi dwarf and standard sized.

Dwarf fruit trees

A dwarf fruit tree is simply a small tree that yields regularly sized fruits. Dwarf trees come in different sizes, from very small ones that you can grow in boxes and pots to larger ones that are almost as tall as the standard sized fruit trees. A dwarf can either be small by nature, or due to external influences. Genetic dwarf fruit trees are very short and have fairly heavy branches due to the nature of their DNA.

Some common varieties of dwarf fruit trees include nectarines, peaches, apples, and apricots that are small enough to grow in pots. These plants usually are very delicate and require winter protection, depending on the climatic conditions in your area.

Semi-dwarf fruit trees

Semi dwarf fruit trees are slightly larger than the dwarfs and can reach up to 12 to 15 feet in size. Once these fruit trees start bearing fruit, you can easily harvest most of the yield without the need of a ladder, with the exception of sweet cherries, as these tend to become a little larger, and can reach up to a height of 15 to 18 feet.

It is interesting to note that an average semi-dwarf tree can yield almost twice as much produce as a dwarf sized fruit tree, without using up more space. Semi-dwarf trees also tend to be well-anchored as compared to their counterpart dwarfs. With proper care and management, you may find a semi dwarf fruit tree perfect for your modest garden or yard, and they can even be grown in containers.

Standard Sized Fruit Trees

The largest sized fruit trees are the standard variety. At their fullest mature size, the standard fruit trees can rise up to heights of 18 to 25+ feet. Depending on the type of tree you want to grow, the mature height will vary just a little. Apricots grow up to a height of 25 to 20 feet, while plums and pears grow up to 18 to 20 feet in height.

The standard sized fruit trees also take longer to yield fruit, but once they start, they produce the highest quantity of all. The challenge is that you may need to use a fruit picker or ladder to help you harvest the fruit at maturity. For spacing purposes, the assumption is that the mature height will be the same as the width of the fruit tree.

As a general rule of thumb, when looking for fruit trees to plant, go for the fruit you enjoy to eat, and then narrow it down to the fruits most suited to your geographical location in order to maximize yields and minimize disease problems.

Fruit Trees and Vines For Tropical Areas

Most tropical fruit trees grow very large and cannot fit well in small backyards, as seen with the large mango trees in Northern Australia that overwhelm small backyards. Avoid growing or buying trees from seeds, as these tend to grow much taller than the grafted ones.

Abiu (Pouteria caimito)

These attractive trees can be kept to three or four meters with regular pruning. They bear bright yellow fruits, about the size of large eggs and are tasty when eaten fresh. Go for grafted trees instead of seedlings, as these can take very long to fruit.

Carambola syn. Five corner fruit

This produces a distinctive yellow fruit and when sliced, it produces a decorative star shape. It has beautiful pink flowers and the tree can grow up to 8 meters in length. You can however keep it to a manageable three feet through regular pruning. These trees can also grow successfully in containers, which helps to limit the size.

Ceylon hill Gooseberry

This very attractive shrub grows to 1.5 m high, but has very ornamental mauve or deep rose pink flowers accompanied with small round purple berries. This plant is frost hardy and tends to fruit generously. It is useful as both a container or hedge plant.

Pawpaw

Homegrown pawpaw trees are the best, but tend to be damaged by frost. You therefore need to grow them in sheltered locations. Normally, male and female flowers are found on separate trees, although there are available bisexual trees. This tree also requires heavy thinning to eliminate the excess males, since the plant requires only one pollinator for every 8 female trees. Female pawpaw flowers have shorter stems than the male flowers, and are attached closer to the tree trunk.

Fruit Trees And Vines For Subtropical Areas

Most subtropical trees tend to grow very large, so you may want to determine the mature size before purchasing.

Acerola syn. Barbados cherry (Malpighia glabra)

This is a very attractive shrub that has dark green glossy leaves and pretty pink flowers. It yields bright red fruits that are extremely rich in vitamin C. Birds are especially fond of this plant, so keep the shrub covered if you decide to grow it.

Grumichana (Eugenia rasiliensis)

This attractive tree with starry white flowers and shiny leaves produces fruits similar to a cherry. It is resistant to fruit fly, but unfortunately popular with the birds. As such, be sure to prune it on a regular basis to keep it bushy enough for a bird net to be thrown over it.

Passion fruit

This is evergreen but can be affected by the woody virus that affects the passion fruit vines. For this reason, be sure to start a new tree every few years. Passion fruit trees are especially fruitful with cross pollination, especially the yellow types.

Strawberry Guava syn. Cherry Guava

This is a stable, evergreen bushy shrub that grows to a height of 3 to 6 meters and whose white flowers blossom in late spring. It has a wine red fruit with a very thin skin. This type of guava is believed to be the most flavorful guava. This tree grows in a variety of soils, but the plant is vulnerable to heavy frost.

Tarmarillo (Cyphomandra betacea)

You can grow these from seeds by getting a suitable fruit from the supermarket and then planting the seeds in spring. Tarmarillos are a short-lived fruit tree, but are very attractive with glossy fruit and huge heart shaped leaves. Normally, orange types are less acidic than the dark red fruits. The added benefit of this fruit tree is that they are resistant to fruit fly.

Fruit Trees and Vines For Temperate Areas

Apples

Apples are one of the greatest fruit trees to grow at home. Apples thrive well in cooler areas and they require a long period of winter cold as well as cross-pollination. Plant only the dwarf apples, because the standard apples can overgrow up to a height of 10 meters or more if not pruned. Ideal apple cultivars to consider include Spartan, Gala, Sturmer Pippin, Akane, Red Fuji and Princess Alexandria. Suitable alternatives are the Polka, Bolero and Waltz as these do not need pruning and can also be grown in pots since they only grow up to a maximum height of 1 to 2 meters. You can either grow these trees as a hedge or in groups as a mini orchard. If you live in a warmer climate, you can grow low chill apples like the Dorset Gold and Anna.

Blueberry (Vaccinium spp.)

These are attractive with small flowers the shape of a bell. They can grow in a wide range of climates, as long as you select the right cultivar. Blueberry trees are selective about soil, and require constant moisture and acid conditions. Ensure that you do not plant the fruit tree near any cement work, and plant more than one for cross-pollination to take place. The Sunshine Blue and Gulf Coast cultivars are a good choice for warmer climates.

*Kiwifruit

These are deciduous vines and require a male and female plant. If you decide to grow this plant, be sure to locate the site in a cooler, east facing, and wind sheltered spot with some winter chill.

*Peach and nectarine

These are great to grow at home, but be sure to select a cultivar suitable to your location. You will also need to do regular pruning to keep the trees productive, and don't forget to protect them from fruit fly and birds. The chilling requirements vary from 150 to 1200 hours. Great Peach cultivar considerations include Glenalton, Starking Delicious, Millicent, Fragar, Halehaven and Anzac. Good cultivar flavors of nectarines include Flavourtop, Independence and Gold mine.

*Plum (plurius spp.)

These should be a definite contender if you are planning to grow 1 or 2 fruit trees, as these are very hardy. There are two main types of plums: European plums (cultivars: President, Grand Duke, Green Gage, and Angelina) and Japanese blood plums (Cultivars: Mariposa, Frontier, Santa Rosa and Satsuma) with each type requiring a different chilling requirement. Japanese plums require 500 to 900 hours and are thus suitable for the warmer, coastal regions, while European plums require 700 to 1000 hrs. All plum trees require cross-pollination, but Japanese plums and European plums will not cross-pollinate. Consider therefore a duo planting or a multi-grafted tree.

Fruit Trees And Vines For Mediterranean Areas

*Figs (Ficus carica)

These can grow in a wide range of climates, but thrive well in areas with dry summers since rain can cause the fruit to split and rot. You can either grow these on a west or north-facing wall.

*Grape (Vitis spp.)

These are useful whenever winter sun and summer shade are required. Since these vines can grow for very many years, be sure to provide a strong pergola or trellis for them to grow on. Disease resistance is especially crucial in coastal or subtropical areas, so good choices include the Carolina Blackrose, Pink lona, Red Flame, Isabella and Muscat Hamburg. Grapes can thrive successfully in pots.

*Mulberry (Morus spp.)

A good mulberry consideration for your small garden is the Shahtoot as this has a cream color and won't stain your clothes. On the other hand, if you would like to plant a black mulberry, take down one on a dwarf rootstock or hunt down a cutting from a good flavored variety. If you plant them closely and prune the trees regularly, you can achieve a fruiting hedge.

*Pomegranate (Punica granatum)

This is a very hardy plant that can withstand extremes of drought, cold and heat. The bronze new growth and vibrant red flowers make this a productive selection for a home garden as well as a highly ornamental choice. Pick the fruit when it is fully ripe.

Quince (Cydonia oblonga)

These are survivors with very beautiful trees and are native to the Middle East. They have lovely pink flowers, and are ideal for the single tree since they need no cross-pollination. You will however need to cook its fragrant golden fruit. Good cultivar considerations include the Smyrna, and Champion.

*Key point/action step

Choosing a suitable site and fruit tree to plant go hand in hand so sometimes, you can choose the kind of tree you want to plant then modify the site to suit that particular fruit. However, sometimes, it is better to determine the characteristics of the site available then choose a fruit tree that can fit within the specifications of the site. Whichever you consider first is okay but the important thing to do is to take your time in this stage since this will determine whether you will harvest fruits or not.

Planting

"Someone is sitting in the shade today because someone planted a tree a long time ago."

- Warren Buffett

By now, you should be decided on the exact type and variety of fruit trees you want to grow in your garden. You will generally find most of the trees you want to grow at your local nursery or garden center. A significant factor to consider at this stage is the timing of your planting i.e. when to plant. Fruit trees tend to do well when sowed in the mid to late fall. It is all right to plant your fruit trees in the early spring, but the best time is usually the late fall. Planting your fruit trees in the fall will give the roots enough time to be established as the ground is still thawed and the air is cold enough to keep the trees from breaking dormancy. All the energy of the trees goes into developing a strong root system beforehand, until the ground freezes. This will ensure a more stable tree with better growth during its first spring as it starts to develop foliage. As a cardinal rule of thumb, most fruit trees will not endure slow drainage soils that keep water for extended periods. Therefore, be sure you are familiar with the soil drainage in your area before planting. Dig a 1-foot deep hole and fill it with water, then wait for 3 to 4 hours to see if the water drains. Fill the hole again, and if the soil takes more than 3 to 4 hours to drain the water during the first and second filling, then you have problems. As mentioned previously, the only three options here are:

*Not to plant there

*Construct a raised bed, mound, or berm and then plant the trees above the current soil

*Install a French drain

The upper part of the trees' root system, the root crown, is the most vulnerable part. In most cases, having your planting area 6 to 12 inches high will be sufficient enough to raise the root crowns above the wet soil. A six-inch mound should be no less than 2 ½ feet in diameter, while a ten to twelve inch berm or mound should be at least 3 to 4 feet wide. To minimize erosion, ensure that the mounds have as much gentle slope as possible.

A good way to ensure that your trees have a higher ground than the surrounding soil is to create a raised bed. For the tastiest fruit and the healthiest trees, go for the sunniest planting location available, except if you are living in a low desert climate where temperatures can reach up to 110 degrees and above in summers.

Spacing will depend on your plan i.e. the number of trees you want to plant, the amount of fruit you want from each, and your intentions on size control. Keep in mind that small trees managed through summer pruning are a lot easier to prune, thin, spray, and harvest, as opposed to larger trees. When planting in high density, separate the trees closely by 18 inches for 2 to 4 trees in one hole, and two or three feet apart for a hedgerow. If there is plenty of space in your garden and you want larger trees, you can just plant them at wider spacing. The choice is yours. You don't need to apply any fertilizer when planting a bare root tree. In fact, fertilizer is not compatible with young, tender feeder roots, and can even kill them along with the tree.

The trees must ultimately grow in the surrounding soil. Avoid making a hole of amended soil at the center of slow draining native soil – water will eventually fill the tree hole and kill the tree. To deal with poorly draining soil, you have to plant in containers or create a raised bed. However, it can be useful to add organic matter such as compost to sandy soil as this will help the soil retain moisture. Consult your local fruit tree nursery for the recommended soil amendments.

The tree should be at the same height as it was in the nursery when you plant it. Be sure not to plant the fruit tree too low. For watering purposes after planting, especially when planting in fast drainage soils, let the trees stand 2 or 3 inches high to accommodate settling. You should plant bare root trees as soon after purchasing as possible. If you buy the trees before the planting day, ensure that the roots are covered or wrapped to maintain high humidity and moisture, then store in a cool place.

How to plant a fruit tree

-Dig a hole that is a little deeper than the length of the root, and make it wide enough to hold the longest roots without having to bend. Loosen the sides of the dug hole. Sometimes roots are unable to penetrate a slick interface.

-Backfill with slightly amended or native soil until you achieve the right planting depth at the bottom of the hole. If you decide to multi plant in one hole, backfill to appropriate depth for every tree.

-Prune off any twisted, broken or rotted roots with a clean cut.

-Place the tree, then spread the roots, and fill up the hole, being sure to tamp the surrounding soil around the roots.

-Water thoroughly if your soil has high drainage in order to complete settling the surrounding soil. Water a little at a time in slower draining soils, over several days if need be. You won't need to water again until the trees grow to several inches high.

How to plant in a raised bed

-For a single tree, create a three to four feet square box, and a 5x5 foot box for 4 trees in one hole.

-Position the box on your poorly draining soil. If necessary, dig a shallow hole for sufficient planting depth.

-Position the tree inside the box, and spread the roots. Fill the box with slightly amended soil or native soil from a different area, tamping the surrounding soil around the roots as you progress. Water the tree as necessary to retain the soil moisture around the roots.

*Key point/action step

Once you have chosen an ideal site and the kind of trees you want to plant, the next important thing is the actual planting. One rule of thumb when planting fruit trees is to ensure that the hole you have dug is large enough to fit the roots without any bending. Additionally, you will need to make sure that you tamp the surrounding soil after filling the hole you had dug with soil after planting. Lastly ensure you water accordingly depending on the soil drainage. For poorly draining soils, you might only need to water once while for soils that drain quickly, you will need to water several times a day for a couple of days.

Pruning and Weeding

"Don't let the tall weeds cast a shadow on the beautiful flowers in your garden."

- Steve Maraboli

Pruning fruit trees is the perfect way to control the size of your fruit trees, stimulate growth, and improve the quality and size of fruit. During pruning, unhealthy, dead or diseased branches are plucked out to give way for new, strong branches and to control the spread of diseases. Fruit trees that have strong, healthy limbs and branches that do not compete or crowd simply have a larger lifespan and are more fruitful than their other counter parts. Moreover, a properly pruned fruit tree is less likely to drop its branches and spread diseases to nearby trees. As a general rule of thumb, any fruit tree should be pruned regularly, regardless of the location or type, in order to ensure the most fruit production.

The best time of the year to prune your fruit trees is the dormant period, between December and mid February. There are two basic types of pruning cuts:

*Heading – This involves removing part of a branch. Choose the right bud when going for a heading cut. The bud normally points in the direction of the new branch.

*Thinning – This involves removing the whole limb or shoot where it originates. Closer cuts are best achieved with scissor type hand pruners. Stubs are more prone to infection.

The terminal bud holds the strongest growth. When cut, the terminal bud is replaced by the lateral bud, and growth progresses in that direction.

Training and pruning systems

*Vase shaped or open center can be used on all nut and fruit trees. It is best for almonds, Asian pears and European plums. This pruning system results in big trees, and it can be a problem to shade from heavy top growth. Choose 3 to 4 limbs evenly distributed around the trunk, in the first year. For sunburn protection and early fruiting, leave small branches on the limbs of these trees. Choose 1 or 2 limbs on each primary in the second year. Head these limbs back to half their lengths, about 24 to 30 inches, then remove the other limbs.

*Central leader results in a small tree, roughly the size of half a vase type, and is brilliant for distributing sunlight. Choose 3 to 5 lateral branches in the first year, lowest to be roughly 12 to 15 inches above the ground, and spaced around the tree evenly, 2 to 3 feet apart vertically. Head the leader and laterals that are more likely to compete with the leader. You may have to physically spread the laterals when they are 5 to 6 feet in length so that they can form a proper angle, ideally 45 degrees with the trunk.

*Y system, like other systems, starts at knee height, and creates a small tree. It is easy to train and is ideal for nectarines and peaches. Trees should be spaced 6 to 7 feet apart, and 15 feet apart in rows. For cherries, pears and plums, increase the distance to 8 – 10 feet between the trees. Cultivate lateral branches from all sides of the "Y".

*Key point/action step

I cannot emphasize enough how important it is to prune your fruit trees. Inability to prune will lead to you having oversized branches that will interfere with the growth of new branches and will eventually affect fruit production. Thus, pruning is a necessity for the best production of fruit.

Pest Control

"Nothing seems to please a fly so much as to be taken for a currant; and if it can be baked in a cake and palmed off on the unwary, it dies happy."

- Mark Twain

Once you have planted your trees and have taken the necessary measures to prune the limbs and branches, the last step before harvesting is to ensure that the trees are pest free. With a little pre planning, it is possible to protect your fruits from major pests with little or no chemical spraying. Practice good husbandry by cultivating the soil around the trees regularly, and collect any leaves and prematurely fallen fruit that may have dropped and destroy them. Inspect the area around your fruit trees regularly and keep it twig free and weed free. It is also important to control aphid infestation and keep other scale insects at bay, even with something as simple as household detergent. Failure to do this will lead to secretion of honeydew from the aphids, which will attract wasps, and subsequently encourage fungus.

*Plum moth caterpillar is most hazardous to gages, damsons and plums. Putting a caterpillar monitor in place in May until early August will help trap the males and control this problem. Tying greasbands around your trees in October can prevent the female moths from climbing the trees and laying their eggs.

*Ladybirds are the best natural pest controls. However, look out for their larva starting May, as these are not as friendly as the adults. Avoid damaging, spraying or crushing them as they will soon become a very significant aphid eater.

*Mice can cause chaos to newly planted fruit trees. The best remedy for this problem is to plant the trees firmly and look out for any new mouse holes around the site during the first growing season. Fortunately, the trees are only vulnerable to these pests for the first year only, after which they become stable enough to repel intruders.

23

*Aphids and caterpillars are come between May and early June. You can control these pests by straightening the leaves once they begin to distort, and then remove by hand.

*Wasps can be very annoying especially to apples and plums. The best approach here is to identify and destroy their nests. However, if you are not up to the task, you can also sacrifice a few plums or apples, cut them into pieces, and drop them at the base of your trees. They usually prefer a readymade meal!

*Peach leaf curl is a condition that pretty much depends on the amount of rainfall in April and your geographical location. The spores responsible for this condition are mainly carried in April rain, therefore the obvious solution is to shield your trees using a slightly weighted, simple polythene sheet in that month.

*Key point/action step

Even if you have planted your fruit trees at an appropriate place and pruned, you also need to ensure that these trees are well protected from pests and diseases. While spraying pesticides in some cases is effective, the chemicals used in pesticides are quite harmful; hence, the reason why I have provided methods of getting rid of pests without having to rely on chemicals.

How to Apply What You've Learned?

You can achieve a generous yield from planting fruit trees on idle land. The good news is that you do not have to plant every year as is the case with grains, but you are still guaranteed of production every year. Fruit trees are also great for rearing bees, which means an added bonus of more honey.

It is important to space the roots of the fruit trees well, and ensure that they are pointing down when you are planting. If the roots get air, the tree can die, so be sure to stamp the soil around them well into the ground.

You can use waste biomass to mulch, and then put rocks at the top. This will help keep the moisture in the soil, and you won't have to water as much. You could also plant lemon grass and comfrey around the pit for future mulch, so that you will need less compost.